CAPE POETRY PAPERBACKS

Margaret Atwood

INTERLUNAR

Margaret Atwood

INTERLUNAR

JONATHAN CAPE
THIRTY-TWO BEDFORD SQUARE
LONDON

First published in Great Britain 1988
Copyright © 1984 by Margaret Atwood
Jonathan Cape Ltd, 32 Bedford Square,
London WC1B 3EL

Some of these poems have appeared in:
*The Canadian Literary Review, Canadian
Literature, Ethos, Saturday Night, This
Magazine, American Poetry Review, Field,
North American Review, River Styx,* and
Poetry Australia. Snake Poems was published
as a limited edition by Salamander Press,
Toronto, in 1983.

A CIP catalogue record for this book
is available from the British Library

ISBN 0-224-02303-9

Printed and bound in Great Britain by
Mackays of Chatham PLC, Chatham, Kent

CONTENTS

CONTENTS

Snake Poems

SNAKE WOMAN

I was once the snake woman,

the only person, it seems, in the whole place
who wasn't terrified of them.

I used to hunt with two sticks
among milkweed and under porches and logs
for this vein of cool green metal
which would run through my fingers like mercury
or turn to a raw bracelet
gripping my wrist:

I could follow them by their odour,
a sick smell, acid and glandular,
part skunk, part inside
of a torn stomach,
the smell of their fear.

Once caught, I'd carry them,
limp and terrorized, into the dining room,
something even men were afraid of.
What fun I had!
Put that thing in my bed and I'll kill you.

Now, I don't know.
Now I'd consider the snake.

LESSON ON SNAKES

Pinned down, this one
opens its mouth as wide as it can
showing fangs and a throat
like the view down a pink lily,
double tongue curved out like stamens.

The lilies do it to keep
from being eaten, this dance of snakes

and the snakes do it to keep from being
eaten also. Since they cannot talk:

the snake is a mute
except for the sound like steam
escaping from a radiator
it makes when cornered:
something punctured and leaking.

This one is green and yellow,
striped like a moose maple.
Sweetly and with grace it hunts
a glimpse, a rustle
among the furry strawberries.

It's hardly
the devil in your garden
but a handy antidote to mice

and yet you'd batter it
with that hoe or crowbar
to a twist of slack rope:

a bad answer
to anything that gets in
what you think is your way.

LIES ABOUT SNAKES

I present the glass snake
which is supposed to break when stepped on
but doesn't. One more lie about snakes,

nor is it transparent. Nothing
could be more opaque. Watch it
there as it undulates over the sand,
a movement of hips in a tight skirt.
You remember the legends
of snakes which were changed to women
and vice versa. Another lie.

Other lies about snakes:
that they cause thunder,
that they won't cross ropes,
that they travel in pairs:
(even when together
for warmth at night or in winter,
snakes are alone)

Swaying up from coiled baskets
they move as if to music,
but snakes cannot hear music.
The time they keep is their own.

BAD MOUTH

There are no leaf-eating snakes.
All are fanged and gorge on blood.
Each one is a hunter's hunter,
nothing more than an endless gullet
pulling itself on over the still-alive prey
like a sock gone ravenous, like an evil glove,
like sheer greed, lithe and devious.

Puff adder buried in hot sand
or poisoning the toes of boots,
for whom killing is easy and careless
as war, as digestion,
why should you be spared?

And you, *Constrictor constrictor*,
sinuous ribbon of true darkness,
one long muscle with eyes and an anus,
looping like thick tar out of the trees
to squeeze the voice from anything edible,
reducing it to scales and belly

And you, pit viper
with your venomous pallid throat
and teeth like syringes
and your nasty radar
homing in on the deep red shadow
nothing else knows it casts . . .
Shall I concede these deaths?

Between us there is no fellow feeling,
as witness: a snake cannot scream.
Observe the alien
chainmail skin, straight out
of science fiction, pure
shiver, pure saturn.

Those who can explain them
can explain anything.

Some say they're a snarled puzzle
only gasoline and a match can untangle.
Even their mating is barely sexual,
a romance between two lengths
of cyanide-coloured string.
Despite their live births and squirming nests
it's hard to believe in snakes loving.

Alone among the animals
the snake does not sing.
The reason for them is the same
as the reason for stars, and not human.

EATING SNAKE

I too have taken the god into my mouth,
chewed it up and tried not to choke on the bones.
Rattlesnake it was, panfried
and good too though a little oily.

(Forget the phallic symbolism:
two differences:
snake tastes like chicken,
and who ever credited the prick with wisdom?)

All peoples are driven
to the point of eating their gods
after a time: it's the old greed
for a plateful of outer space, that craving for darkness,
the lust to feel what it does to you
when your teeth meet in divinity, in the flesh,
when you swallow it down
and you can see with its own cold eyes,
look out through murder.

This is a lot of fuss to make about mere lunch:
metaphysics with onions.
The snake was not served with its tail in its mouth
as would have been appropriate.
Instead the cook nailed the skin to the wall,
complete with rattles, and the head was mounted.
It was only a snake after all.

(Nevertheless, the authorities are agreed:
God is round.)

METEMPSYCHOSIS

Somebody's grandmother glides through the bracken,
in widow's black and graceful
and sharp as ever: see how her eyes glitter!

Who were you when you were a snake?

This one was a dancer who is now
a green streamer waved by its own breeze
and here's your blunt striped uncle, come back
to bask under the wicker chairs
on the porch and watch over you.

Unfurling itself from its cast skin,
the snake proclaims resurrection
to all believers

though some tire soon of being born
over and over; for them there's the breath
that shivers in the yellow grass,
a papery finger, half of a noose, a summons
to the dead river.

Who's that in the cold cellar
with the apples and the rats? Whose is
that voice of a husk rasping in the wind?
Your lost child whispering *Mother*,
the one more child you never had,
your child who wants back in.

THE WHITE SNAKE

The white snake is to be found, says legend,
at the dark of the moon,
by the forks of roads, under three-leaved trees,
at the bottoms of unsounded lakes.

It looks like water
freezing. It has no eyes.
It lays quartz eggs and foretells the future.

If you can find it and eat it
then you will understand
the languages of the animals.

There was a man who tried it.
He hunted, caught, transformed
the sacred body of living snow
into raw meat, cut into it, swallowed.

Then sound poured over him
like a wall breaking, like a disaster:

He went blind in an instant.
Light rose in him
filling his mouth like blood,
like earth in the mouth of a man buried.

Human speech left him.
For the rest of his life, emptied and mute
he could do nothing but listen
to the words, words around him everywhere like rain falling.

Beware of the white snake, says the story.
Choose ignorance.

(There are no white snakes in nature.)

PSALM TO SNAKE

O snake, you are an argument
for poetry:

a shift among dry leaves
when there is no wind,
a thin line moving through

that which is not
time, creating time,
a voice from the dead, oblique

and silent. A movement
from left to right,
a vanishing. Prophet under a stone.

I know you're there
even when I can't see you

I see the trail you make
in the blank sand, in the morning

I see the point
of intersection, the whiplash
across the eye. I see the kill.

O long word, cold-blooded and perfect

QUATTROCENTO

The snake enters your dreams through paintings:
this one, of a formal garden
in which there are always three:

the thin man with the green-white skin
that marks him vegetarian
and the woman with a swayback and hard breasts
that look stuck on

and the snake, vertical and with a head
that's face-coloured and haired like a woman's.

Everyone looks unhappy,
even the few zoo animals, stippled with sun,
even the angel who's like a slab
of flaming laundry, hovering
up there with his sword of fire,
unable as yet to strike.

There's no love here.
Maybe it's the boredom.

And that's no apple but a heart
torn out of someone
in this myth gone suddenly Aztec.

This is the possibility of death
the snake is offering:
death upon death squeezed together,
a blood snowball.

To devour it is to fall out
of the still unending noon
to a hard ground with a straight horizon

and you are no longer the
idea of a body but a body,
you slide down into your body as into hot mud.

You feel the membranes of disease
close over your head, and history
occurs to you and space enfolds
you in its armies, in its nights, and you
must learn to see in darkness.

Here you can praise the light,
having so little of it:

it's the death you carry in you
red and captured, that makes the world
shine for you
as it never did before.

This is how you learn prayer.

Love is choosing, the snake said.
The kingdom of god is within you
because you ate it.

AFTER HERACLITUS

The snake is one name of God,
my teacher said:
All nature is a fire
we burn in and are
renewed, one skin
shed and then another.

To talk with the body
is what the snake does, letter
after letter formed on the grass,
itself a tongue, looping its earthy hieroglyphs,
the sunlight praising it
as it shines there on the doorstep,
a green light blessing your house.

This is the voice
you could pray to for the answers
to your sickness:
leave it a bowl of milk,
watch it drink

You do not pray, but go for the shovel,
old blood on the blade

But pick it up and you would hold
the darkness that you fear
turned flesh and embers,
cool power coiling into your wrists
and it would be in your hands
where it always has been.

This is the nameless one
giving itself a name,
one among many

and your own name as well.

You know this and still kill it.

THE BLUE SNAKE

The snake winds through your head
into the temple which stands on a hill
and is not much visited now.

Toppled stones clutter the paving
where the blue snake swims towards you,
dry in the dry air,
blue as a vein or a fading bruise.
It looks at you from the side of its head
as snakes do. It flickers.

What does it know
that it needs to tell you?
What do you need to be told?

You are surprised to hear it speak.
It has the voice of a flute
when you first blow into it,
long and breathless; it has an old voice,
like the blue stars, like the unborn,
the voice of things beginning and ceasing.

As you listen, you grow heavier.
It asks you why you are here,
and you can't answer.

It begins to glow,
it's almost transparent now,
you can see the spine
with its many pairs of delicate ribs
unrolling like a feather.

This has gone far enough,
you think, and turn away.
It isn't what you came for.

Behind you the snake dissolves
and flows into the rock.

On the plain below you is a river
you know you must follow home.

Interlunar

I

DOORWAY

I seem to myself to be without power.
To have the power of waiting merely.
Waiting to be told what to say.
But who will tell me?

November is the month of entrance,
month of descent. Which has passed easily,
which has been lenient with me this year.
Nobody's blood on the floor.

My arm lies across this oak desk
in the fading sunlight of four o'clock,
the skin warming, alive still,
the hand unspoken.

Through the window
below the half-lowered blind, there are
the herbs frost-killed in their boxes,
life retreating to the roots;
beyond them, the rubbishy laneway
owned by nobody.
Where all power is either spent or potential.

Power of the grey stone
resting inert, not shaping itself.
Power of the murdered girl's
bone in the stream, not yet a flute.
Power of a door unopened.

BEFORE
For John T.

A bowl of cooked earth, holding
itself in another form:
the last apples of this winter,
wizened and sticky.
Possibly not food.

Even the weak spring sunlight
waning through the four small panes
is excess for him:
his eyes slit against it.

A knife, left over from last night,
eases to the surface, turns over
lazily, floats there,
a fish rising
in brown swamp water,
silver and dead.

Some woman or other,
damp handfuls of flesh
and no questions.
He is fat bait.

Desire gathers
icy and glistening, along
the buds of the forsythia
about to open.

Who cares who cares.
He would rather be a black stone
in the dirt back of the garden
or God. The same thing.

The hoe and the mattock lean in the corner.

Out there,
along the edge of the saltmarsh
two crows level, their wings
radiating darkness.

It's only one version.
So is the sun.

BEDSIDE

You sit beside the bed
in the *extremis* ward, holding your father's feet
as you have not done since you were a child.
You would hold his hands, but they are strapped down,
emptied at last of power.

He can see, possibly, the weave of the sheet
that covers him from chest to ankles;
he does not wish to.

He has been opened. He is at the mercy.

You hold his feet,
not moving. You would like
to drag him back. You remember
how you have judged each other
in silence, relentlessly.

You listen intently, as if for a signal,
to the undersea ping of the monitors,
the waterlogged lungs breathed into by machines,
the heart, wired for sound
and running too quickly in the stuck body,

the murderous body, the body
itself stalled in a field of ice
that spreads out endlessly under it,
the snowdrifts tucked by the wind around
the limbs and torso.

Now he is walking
somewhere you cannot follow,
leaving no footprints.
Already in this whiteness
he casts no shadow.

A HOLIDAY

My child in the smoke of the fire
playing at barbarism,
the burst meat dripping down her
chin, soot smearing
her cheek and her hair infested with twigs,
under a huge midsummer-leafed tree
in the rain, the shelter
of poles and canvas down
the road if needed:

This could be where we
end up, learning the minimal
with maybe no tree, no rain,
no shelter, no roast carcasses
of animals to renew us

at a time when language
will shrink to the word *hunger*
and the word *none*.

Mist lifts from the warm lake
hit by the cold drizzle:
too much dust in the stratosphere
this year, they say. Unseasonal.

Here comes the ice,
here comes something,
we can all feel it
like a breath, a footstep,
here comes nothing
with its calm eye of fire.

What we're having right
now is a cookout,
sausages on peeled sticks.
The blades of grass are still with us.
My daughter forages,
grace plumps the dusty berries,
two or three hot and squashed in her fist.

So far we do it
for fun. So far is
where we've gone
and no farther.

LUNCHTIME DURING A PEAK YEAR
IN THE YELLOWJACKET CYCLE

Right now we live in tents
and wake in the orange light and tar
smell of canvas heating
in the sun. When it rains we play cards.
We eat our foods
in the order in which they will otherwise
rot, the hardiest last.

Today the lake simmers,
bright as the tin plates
we wash in it. The wasps are thick,
lured by the chance of perishing
and spoilage; they flounder
in the enamel cups of smoky
tea, perch on the damp paper
the meat was wrapped in before it melted,
the peaches in sweet syrup
tasting of metal,
the spoons raised halfway to our mouths.
They saw and gorge themselves and stagger
into the air, dizzy with blood and sugar.

The stew muscle in its juices
ferments over the fire.
A wasp falls into the pot,
flutters like a tinsel heart
and is cooked, adding its tiny hurt.

Look before you eat, says my mother
as she stoops and ladles.
Her nomadic children ignore her.
Every year she is shorter '
and we are more oblivious.
Already we are beyond her power
to save us from even small disasters.
We have our own purposes.
We think we can do what we want.

This is the summer I am going to devour
everything I can dig up or strangle,
each muddy clam and bitter stem,
looking it up in the book first.

I squat, the gold wasps crawling
harmless over my bare arms and leathery
toes, thinking about my secret meals,
nearly savage and single-
minded and living
off the land, stirring a purple
swampy cattail soup
in a tomato can on a hot stone
and burning it, ferocious
with hunger for every untried
food, dazed by the sunlight
and abundance, knowing nothing about death.

THE HEALER

I do not wish to spend the rest of my time
curing nosebleeds over the phone. Nobody here
needs anyone raised from the dead, it's too
confusing, with the notices
already sent and so forth. Asthma and bruises, warts
dissolved in moonlight, a touch and it's done.
Anything worse and they'll call the doctor.
Suffering is boring,
though noticing this does not make it end.
There are so many other things I could be doing
with my hands: digging up the garden,
digging up the garden again.
Sometimes I think my life is over and there will be
repetition but no more story.
Only these compassions, which are also minor.

I should be elsewhere, away from these
neat farms, living among the people in dust-floored
shacks who could still believe me.
But I am old and lazy now. I know that
being sick and being well are states
of the soul, though I am losing
ground, call it altitude, call it faith.
The power is in me, but what for?
What am I to do with my hands in this tidy place
filled with those who do not want
to be truly healed?

Such arrogance, to have expected miracles.
What was it anyway I thought flowed through me?
Perhaps it was only a slight talent, this tinkering
with the small breaks and fissures in other bodies,
like a knack for crewel-work.

Sundays I putter in the yard, àrranging stones,
raking grass, and the church-goers pass me,
radiating their special hatreds.
In the evenings I sit on the back porch
in a stuffed chair covered with blue cloth
printed with flowers, and look out
across the ragged fields at the real
flowers, goldenrod and purple asters,
the light spilling out of them
unasked for and unused.

THE SAINTS

The saints cannot distinguish
between being with other people and being
alone: another good reason for becoming one.

They live in trees and eat air.
Staring past or through us, they see
things which we would call not there.
We on the contrary see them.

They smell of old fur coats
stored for a long time in the attic.
When they move they ripple.
Two of them passed here yesterday,
filled and vacated and filled
by the wind, like drained pillows
blowing across a derelict lot,
their twisted and scorched feet
not touching the ground,
their feathers catching in thistles.
What they touched emptied of colour.

Whether they are dead or not
is a moot point.
Shreds of them litter history,
a hand here, a bone there:
is it suffering or goodness
that makes them holy,
or can anyone tell the difference?

Though they pray, they do not pray
for us. Prayers peel off them
like burned skin healing.
Once they tried to save something,
others or their own souls.
Now they seem to have no use,
like the colours on blind fish.
Nevertheless they are sacred.

They drift through the atmosphere,
their blue eyes sucked dry
by the ordeal of seeing,
exuding gaps in the landscape as water
exudes mist. They blink
and reality shivers.

NOMADS

A woman speaking of a man
thinking she means comfort
when what in fact she means is
blood in the bathtub.
Or it could be a man speaking
about a woman.

The nomads file past us
in the forests
at night, their eyes picking up
the minor light, like an owl's,
going their rounds. Going
home, which is motion.
The children sleeping on the warm backs
among the fur and hair.
The craving for arms around
us; the belief we need
that the trees love us.

He's so good, she says,
knowing already the hole *No* makes
in the center of her forehead
where the eye was once that could
see it coming.
The outline of him stands
over there, against
the wall, devoid of features.
She should know by now
the disguises your own death takes
when it really wants you.

VALEDICTION, INTERGALACTIC

Now that the pain is slower I know
it's there, less
like being flayed than being
scalded. A long moment of no breath at all
and no feeling. Then layer after layer peels off.
A peach in boiling water.
This is a domestic image.
Try: soft moon with the rind off.
The more I go on the less it's
anyone's fault, especially not
yours, who got neediness done to you
decades ago, and the doctor
doesn't stop to ask why your
blood and eggwhite is coming
out over the floor but shoves it back in
and calls for a suture. Which is what
I'm doing, though all that mending to keep things
together and smooth ruined my eyes

so at the end I could see only
the shift between light and dark, and you were
light at first and then dark
and then light and then dark, and I wanted it
to be light all the time, as in religious
postcards, or the arctic circle.
Is this intolerance? Am I
non-human? Is it greed for some
stupid absolute, some zero,
that takes my skin off
like this and makes your unsaid
words flare blue with terror? Do I prefer
the airless blaze of outer
space to men, even
the beautiful ones? Goodbye
earthling, you were more perfect
than anyone, though far from it.

PRECOGNITION

Living backwards means only
I must suffer everything twice.
Those picnics were already loss:
with the dragonflies and the clear streams halfway.

What good did it do me to know
how far along you would come with me
and when you would return?
By yourself, to a life you call daily.

You did not consider me a soul
but a landscape, not even one
I recognize as mine, but foreign
and rich in curios:
an egg of blue marble,
a dried pod,
a clay goddess you picked up at a stall
somewhere among the dun and dust-green
hills and the bronze-hot
sun and the odd shadows,

not knowing what would be protection,
or even the need for it then.

I wake in the early dawn and there is the roadway
shattered, and the glass and blood,
from an intersection that has happened
already, though I can't say when.
Simply that it will happen.

What could I tell you now that would keep you
safe or warn you?
What good would it do?
Live and be happy.

I would rather cut myself loose
from time, shave off my hair
and stand at a crossroads
with a wooden bowl, throwing
myself on the dubious mercy
of the present, which is innocent
and forgetful and hits the eye bare

and without words and without even love
than do this mourning over.

HIDDEN

In the warm dusk over the exhausted trees
outside the farmhouse, in the scent
of soft brick and hot tin,

weeding among the white-pink peonies,
their sugary heads heavy and swollen,
I can hear them breathing out
and then out again, as if giving up.

It's June, the month when the dead
are least active though most hungry

and I'm too close to the ground, to those
who have faded and merged, too close
to contagion. Hidden in the border somewhere
near, a bone sings of betrayal.

My fingers are wet with bruised
green stems and dewfall.
In this season of opening out,
there is something I want closed.

The sun sinks and the body darkens
from within: I can see the light going out of my hands.

I doubt that I ever loved you.
I believe I have chosen peace.

Think of this as the dormant phase
of a disease.

KEEP

I know that you will die
before I do.

Already your skin tastes faintly
of the acid that is eating through you.

None of this, none of this is true,
no more than a leaf is botany,

along this avenue of old maples
the birds fall down through the branches
as the long slow rain of small bodies
falls like snow through the darkening sea,

wet things in turn move up out of the earth,
your body is liquid in my hands, almost
a piece of solid water.

Time is what we're doing,
I'm falling into the flesh,
into the sadness of the body
that cannot give up its habits,
habits of the hands and skin.

I will be one of those old women
with good bones and stringy necks
who will not let go of anything.

You'll be there. You'll keep
your distance,
the same one.

ANCHORAGE

This is the sea then, once·
again, warm this time
and swarming. Sores fester
on your feet in the tepid
beach water, where French
wine bottles float among grape-
fruit peels and the stench of death
from the piles of sucked-out shells
and emptied lunches.
Here is a pool with nurse sharks
kept for the tourists
and sea turtles scummy with algae,
winging their way through their closed
heaven of dirty stones. Here
is where the good ship *Envious*
rides at anchor.
The land is red with hibiscus
and smells of piss; and here
beside the houses built on stilts,
warped in the salt and heat,
they plant their fathers in the yards,
cover them with cement
tender as blankets:

Drowned at sea, the same one
the mermaids swim in, hairy
and pallid, with rum on the beach after.
But that's a day trip.
Further along, there are tents
where the fishers camp,
cooking their stews of claws
and spines, and at dawn they steer
further out than you'd think
possible, between the killer
water and the killer sun,
carried on hollow pieces
of wood with the names of women,
not sweethearts
only but mothers, clumsy
and matronly, though their ribbed bodies
are fragile as real bodies
and like them also a memory,
and like them also two hands
held open, and like them also
the last hope of safety.

GEORGIA BEACH

In winter the beach is empty
but south, so there is no snow.

Empty can mean either
peaceful or desolate.

Two kinds of people walk here:
those who think they have love
and those who think they are without it.

I am neither one nor the other.

I pick up the vacant shells,
for which *open* means *killed*,
saving only the most perfect,
not knowing who they are for.

Near the water there are skinless
trees, fluid, greyed by weather,
in shapes of agony, or you could say
grace or passion as easily.
In any case twisted.

By the wind, which keeps going.
The empty space, which is not empty
space, moves through me.

I come back past the salt marsh,
dull yellow and rust-coloured,
which whispers to itself,
which is sad only to us.

A SUNDAY DRIVE

The skin seethes in the heat
which roars out from the sun, wave after tidal wave;
the sea is flat and hot and too bright,
stagnant as a puddle,
edged by a beach reeking of shit.
The city is like a city
bombed out and burning;
the smell of smoke is everywhere,
drifting from the mounds of rubble.
Now and then a new tower,
already stained, lifts from the tangle;
the cars stall and bellow.
From the trampled earth rubbish erupts
and huts of tin and warped boards
and cloth and anything scavenged.
Everything is the colour of dirt
except the kites, red and purple,
three of them, fluttering cheerfully
from a slope of garbage,
and the womens' dresses, cleaned somehow,
vaporous and brilliant, and the dutiful
white smiles of the child beggars
who kiss your small change
and press it to their heads and hearts.

Uncle, they call you. *Mother*.
I have never felt less motherly.
The moon is responsible for all this,
goddess of increase
and death, which here are the same.
Why try to redeem

anything? In this maze
of condemned flesh without beginning or end
where the pulp of the body steams and bloats
and spawns and multiplies itself
the wise man chooses serenity.

Here you are taught the need to be holy,
to wash a lot and live apart.
Burial by fire is the last mercy:
decay is reserved for the living.

The desire to be loved is the last illusion:
Give it up and you will be free.

Bombay, 1982

II

ONE SPECIES OF LOVE,
AFTER A PAINTING
BY HIERONYMOUS BOSCH

In the foreground there are a lot of stones,
each one painted singly
and in detail.

There is a man, sitting down.
Behind the man is a hill,
shaped like a mound burial .
or a pudding,
with scrubby bushes, the leaves glazed
by the serene eye-colour of the sky.
In the middle distance, an invisible line
beyond which things become
abruptly bluer.

At the man's feet there is a lion,
plush-furred and blunted,
and in the right foreground, a creature
part bird, part teapot,
part lizard and part hat
is coming out of an eggshell.

The man himself, in his robe
the muted pink of the ends of fingers
is gazing up at the half-sized
woman who is suspended
in the air over his head.

She has wings, but they aren't moving.
She's blue, like the background,
denoting holiness or distance
or perhaps lack of a body.

She holds one hand in a gesture
of benediction which is a little wooden.
The other hand points to the ground.

There is no sound in this picture,
light but no shadows.
The stones keep still.
The surface is clear
and without texture.

GISELLE IN DAYTIME

You know the landscape: in the distance
three low hills, bare of snow.

In the foreground the willow grove
along the unmoving river
not icebound. Snow in the shadows though.
A diffused light that is not the sun.

Here and there, the young girls
in their white dresses made of paper
not written on.
No one is here willingly.
None are mourners.

Each stays under a separate tree,
sitting or standing as if
aimless. It was not
an end they wanted but more life.

The near one crouches on the chilled
sand, knees to belly,
holding in her hands a plain stone
she turns over and over,
puzzled, searching for the cut in it
where the blood ran out.

The tree arching above her
is dead, like everything
here. Nevertheless it sways, although there is
no wind, quickening and shaking out
for her its thin leaves and small green flowers,

which has never happened before,
which happens every day,
which she does not notice.

ORPHEUS (1)

You walked in front of me,
pulling me back out
to the green light that had once
grown fangs and killed me.

I was obedient, but
numb, like an arm
gone to sleep; the return
to time was not my choice.

By then I was used to silence.
Though something stretched between us
like a whisper, like a rope:
my former name,
drawn tight.
You had your old leash
with you, love you might call it,
and your flesh voice.

Before your eyes you held steady
the image of what you wanted
me to become: living again.
It was this hope of yours that kept me following.

I was your hallucination, listening
and floral, and you were singing me:
already new skin was forming on me
within the luminous misty shroud
of my other body; already
there was dirt on my hands and I was thirsty.

I could see only the outline
of your head and shoulders,
black against the cave mouth,
and so could not see your face
at all, when you turned

and called to me because you had
already lost me. The last
I saw of you was a dark oval.
Though I knew how this failure
would hurt you, I had to
fold like a grey moth and let go.

You could not believe I was more than your echo.

EURYDICE

He is here, come down to look for you.
It is the song that calls you back,
a song of joy and suffering
equally: a promise:
that things will be different up there
than they were last time.

You would rather have gone on feeling nothing,
emptiness and silence; the stagnant peace
of the deepest sea, which is easier
than the noise and flesh of the surface.

You are used to these blanched dim corridors,
you are used to the king
who passes you without speaking.

The other one is different
and you almost remember him.
He says he is singing to you
because he loves you,

not as you are now,
so chilled and minimal: moving and still
both, like a white curtain blowing
in the draft from a half-opened window
beside a chair on which nobody sits.

He wants you to be what he calls real.
He wants you to stop light.
He wants to feel himself thickening
like a treetrunk or a haunch
and see blood on his eyelids
when he closes them, and the sun beating.

This love of his is not something
he can do if you aren't there,
but what you knew suddenly as you left your body
cooling and whitening on the lawn

was that you love him anywhere,
even in this land of no memory,
even in this domain of hunger.
You hold love in your hand, a red seed
you had forgotten you were holding.

He has come almost too far.
He cannot believe without seeing,
and it's dark here.
Go back, you whisper,

but he wants to be fed again
by you. O handful of gauze, little
bandage, handful of cold
air, it is not through him
you will get your freedom.

THE ROBBER BRIDEGROOM

He would like not to kill. He would like
what he imagines other men have,
instead of this red compulsion. Why do the women
fail him and die badly? He would like to kill them gently,
finger by finger and with great tenderness, so that
at the end they would melt into him
with gratitude for his skill and the final pleasure
he still believes he could bring them
if only they would accept him,
but they scream too much and make him angry.
Then he goes for the soul, rummaging
in their flesh for it, despotic with self-pity,
hunting among the nerves and the shards
of their faces for the one thing
he needs to live, and lost
back there in the poplar and spruce forest
in the watery moonlight, where his young bride,
pale but only a little frightened,
her hands glimmering with his own approaching
death, gropes her way towards him
along the obscure path, from white stone
to white stone, ignorant and singing,
dreaming of him as he is.

LETTER FROM PERSEPHONE

This is for the left-handed mothers
in their fringed black shawls or flowered housecoats
of the 'forties, their pink mule slippers,
their fingers, painted red or splay-knuckled
that played the piano formerly.

I know about your houseplants
that always died, about your spread
thighs roped down and split
between, and afterwards
that struggle of amputees
under a hospital sheet that passed
for sex and was never mentioned,
your invalid mothers, your boredom,
the enraged sheen of your floors;
I know about your fathers
who wanted sons.

These are the sons
you pronounced with your bodies,
the only words you could
be expected to say,
these flesh stutters.

No wonder this one
is nearly mute, flinches when touched,
is afraid of caves
and this one threw himself at a train
so he could feel his own heartbeat
once anyway; and this one
touched his own baby gently
he thought, and it came undone;
and this one enters the trussed bodies
of women as if spitting.

I know you cry at night
and they do, and they are looking for you.

They wash up here, I get
this piece or that. It's a blood
puzzle.

It's not your fault
either, but I can't fix it.

THREE DENIZEN SONGS

I

Everyone is afraid of me.
Why is that?
Is it the history that shines on my skin
and fits me like satin?

Is it my potential, the energy
of an open socket, a dark
vortex in the wall, which is never seen
doing anything?

My eyes are pure archeology,
through which you can see straight down
past all those bones and broken
kitchen utensils and slaughters,

yet it's my happiness they envy,
happiness of a lizard in the sun,
a mailed spiral
gorging on light, not bothering anyone.

Whenever I go to the window
fifty people yell *jump* at me
silently. They like
accidents in this country,
the worse the better.
Fatal diseases will also do.

If I had a really bad accident
they would not be afraid of me any longer.

If I had a really bad accident,
would you love me?

II

I advance towards you along the street,
one foot in front of the other,
cotton skirt billowing,
a child twirling at the end of each arm

and you see me not as human
but as cavern:
larval darkness and velvet shelter,
a rural motif, like a cow-fleshed
pumpkin, redolent and milky in the barn,
or an oriole's nest, pendulous and hairy

or: converse:
a vacancy pumping night out
through the ends of my fingers, a cleft heart,
drainhole and quagmire, or a lagoon
at night, black and oily,
out of which something venusian
with skinfolds and many teeth will certainly lift soon

or a lush moon, sensuous
and damp as new mushrooms, or a meat egg,
salmon-coloured; in any case
curved space.

Is it my body or your vision
which is martian?

III

(After Ray Bradbury)

All suspicions are justified
sooner or later.
When after the chase I'm finally cornered
I'll lie there on the sidewalk
the colours racing across my body
as if I were a dying squid
or a slug with salt on it
turning to irridescent gruel

while they insert their curious probes
into my third eye to see
what kind of creature I am
really, and prove me alien,
and scientists come running
on the double to prevent needless cruelty
and small boys gather stones.

I'm no Martian
I'll cry in gutteral,
in distress, waving my invisible tentacles:
I'm only your little sister
who died before you were born . . .

They won't believe me;
they know I'm not human,
despite this pink print
dress; I'm only everyone's fantasy
of loss. Under
duress (and what else is there)
I know I'll say
anything. Anything.

SINGING TO GENGHIS KHAN

In the plum-coloured tent in the evening
a young woman is playing a lute,
an anachronism,
and singing to Genghis Khan.

It is her job. It is her intention
to make him feel better.
Then maybe she can get some sleep
and will not be murdered.

He sits up nights, brooding,
which makes him bad-tempered.
He thinks the sky is a black void.
He thinks he is made of dust.
He thinks he suffers.

The woman plucks the strings
and keeps on singing,
arranging her lips gracefully
and feeling the silk of her dress
against her skin.
Most of the things he thinks about
are of no importance to her.
She sings about gardens.

Listening, he hugs himself tighter
in his mantle, his exile.
No one can take it from him.
Women, he thinks, are exiles
only from men, but men are
exiles from everything.
He is cast out from her.
He equates her with Nature:
she does not care.

The singing woman is sorry for him.
Also she thinks he is stupid,
but does not say so.
Also she is afraid of him,
and his despair, his monster.
She has no wish to be food.

After gardens she tries the sea,
then birds, then battles.
Then she offers him
her flesh to pacify him.
In the oily light he turns her
this way and that,

But he will not be consoled
by her or pleasure or any
thing, for being
alive on this earth,
for all the women he can never
throw himself into,
for being locked in his body
which will get bigger
or fly, for the thousands of men
he can never kill. For the verb *to die*.

HARVEST

The villagers are out hunting for you.
They have had enough of potions
for obtaining love: they do not want love
this year, the crops were scant, a bad wind
came with the fall mists and they want you.

Already they have burned your house,
broken your mirror
in which they used to glimpse over their shoulders
the crescent moon and the face
of the one desired,
kicked the charred bedding apart
looking for amulets, gutted your cats.

By night you walk in the fields, fending off
the voices that rustle in the air
around you, wading streams
to kill your smell,
or creep into the barns to steal milk
and the turnips the pigs eat.
By day you hide,
digging yourself in under the hedges,
your dress becoming the colour of ashes.
Praying to the rain to save you.

Through the smooth grey tree-trunks there are fragments
of coats, the red wool
sashes, whistles rallying the dogs,
drifting towards you through the leaves falling
like snow, like pestilence.

Of the men who will stand around
the peeled stake dug into a pit,
bundles of sticks and dried reeds piled
nearby, the mud caked on their bootsoles,
crooking their fingers to ward you off,
at the same time joking about
your slashed breasts and the avid
heat to come, there is not one

who has not run his hands over your skin
in stealth, yours or your shadow's,
not one who has not straddled you
in the turned furrows, begging for increase.

What you will see: the sun, for the last time.
The tree-strewn landscape gathering itself
around you; the blighted meadows.
The dark ring of men whose names
you know, and their bodies
you remember as nameless, luminous.
The children, on the edges

of the circle, twisting on the ground,
arms uplifted and spread, mouths open, playing
at being you. In the distance, the wives
in their flowered shawls and thick decorous
skirts, hurrying from their houses
with little pierced copper
bowls in their hands, as if offering food
at a feast, bringing the embers.

A MASSACRE BEFORE
IT IS HEARD ABOUT

Sometimes there is an idea
so pure it is without mercy.

It sweeps over the wet fields
where rice is planted by women
in bare feet and pink skirts
and hits the jungles like a blade
of hard light. Love
in the abstract is deadly.

Some died for embracing
each other, some for speaking,
some for remaining silent,
some for remaining.
The wells filled up. After a time
nobody bothered burying.

In a jungle this hot
the afterlife happens quickly
and tenderly. Ants flow like water.
Week-old bodies
are already without flesh and eyes.
Even so many of them.
Tendrils grow around them
as if forgiving them.
Vines break out their raucous flowers
despite them.

The ruinous huts, the parts
of children gnawed by cats, the cooking
fires left smouldering, the cairns
of bones arranged so neatly.
In the service
of the word.

I would say the stones
cry out, except they don't.
Nothing cries out. The light falls on all of this
equally. Fear and memory
work their way down into the earth
and lie fallow.

NO NAME

This is the nightmare you now have frequently:
that a man will come to your house at evening
with a hole in him—you place it
in the chest, on the left side—and blood leaking out
onto the wooden door as he leans against it.

He is a man in the act of vanishing
one way or another.
He wants you to let him in.
He is like the soul of a dead
lover, come back to the surface of the earth
because he did not have enough of it and is still hungry

but he is far from dead. Though the hair
lifts on your arms and cold
air flows over your threshold
from him, you have never
seen anyone so alive

as he touches, just touches your hand
with his left hand, the clean
one, and whispers *Please*
in any language.

You are not a doctor or anything like it.
You have led a plain life
which anyone looking would call blameless.
On the table behind you
there are bread on a plate, fruit in a bowl.
There is one knife. There is one chair.

It is spring, and the night wind
is moist with the smell of turned loam
and the early flowers;
the moon pours out its beauty
which you see as beauty finally,
warm and offering everything.
You have only to take.
In the distance you hear dogs barking.

Your door is either half open
or half closed.
It stays that way and you cannot wake.

LETTER FROM THE
HOUSE OF QUESTIONS

Everything about me is broken.
Even my fingers, forming
these words in the dust
a bootprint will wipe out by morning,
even these words. Even, almost,
my will to do anything.
I would spend the rest of my life
in a house corner, in the sun.
If there were a house. If there were sun.

Those who imagine our bodies
for their own pleasure, in their beds
that smell of mouthwash and lotion
have never been here or anywhere like it.
We stink. Our clothes and breath
and the pissed floor and air
and the wounds they picture as the stigmata
of desire, which are just holes
dug in the flesh and filthy
with neglect. Blunt and vicious,
nothing to do with the arabesques
of civilized lust they invent.

Lady, I saw you once
in a photograph in a magazine
before I was brought here.
You were in a kitchen, with children
seated at a table. You stood,
serving red and yellow and green food
and holding up a bottle.
Everyone smiled. I have forgotten
what you were supposed to be
selling, but I remember you,
with your white teeth shining in the room
which was so clean and beautiful.

I lie on the floor and hold on to you
as if you were here. As if you were
a picture of myself
when I still existed.
Look down on me. Explain to me
why this is happening. Send food.
The saints having failed me,
you are the one I pray to.

ORPHEUS (2)

Whether he will go on singing
or not, knowing what he knows
of the horror of this world:

He was not wandering among meadows
all this time. He was down there
among the mouthless ones, among
those with no fingers, those
whose names are forbidden,
those washed up eaten into
among the grey stones
of the shore where nobody goes
through fear. Those with silence.

He has been trying to sing
love into existence again
and he has failed.

Yet he will continue
to sing, in the stadium
crowded with the already dead
who raise their eyeless faces
to listen to him; while the red flowers
grow up and splatter open
against the walls.

They have cut off both his hands
and soon they will tear
his head from his body in one burst
of furious refusal.
He foresees this. Yet he will go on
singing, and in praise.
To sing is either praise
or defiance. Praise is defiance.

READING A POLITICAL THRILLER
BESIDE A REMOTE LAKE
IN THE CANADIAN SHIELD

I give you yourself, or
me, back propped
against a rock, with a skin of brown lichen
edible in extreme need.
This rock came straight up out of the earth
which was not earth then.
It is much older
than these words, or the hands
that hold them here, too close
to your eyes, this paper
world which is the real world
also: intricate wire
devices, lies, men who pretend
to kill because they must,
blowing each other up
in the best of causes,
and the refugees who die and refuse
to forget, and die once more.
Almost you envy
their hatred and ferocity,
which at least make everything clear,
narrow a landscape down
to a blunt point: target and power.

Is this plot more important
than the sunset?
Which is thick as blood but nothing
of the kind:
tonight flamingo, melting
into the water like love, true
and fervent: motif
hackneyed as bravery,
which criminals practise also.

Is it viciousness of the genes
that drives us on,
the quest for protein?
Starvation too is a creed.

There's a moon. What could be purer?
In this light, hard to believe
in any motive.
The trees go on and on.
Granite against your flesh, your flesh
empty and peaceful. Too dark to read.

In the closed book, the rivals
are at each others' throats
for a few miles of stone, a village
you wouldn't look twice at.

For them it is not a story.
This lake is dying.
Light the lamp,
pull the forest up over your body.
Sleep while you can.

III

THE WORDS CONTINUE
THEIR JOURNEY

Do poets really suffer more
than other people? Isn't it only
that they get their pictures taken
and are seen to do it?
The loony bins are full of those
who never wrote a poem.
Most suicides are not
poets: a good statistic.

Some days though I want, still,
to be like other people;
but then I go and talk with them,
these people who are supposed to be
other, and they are much like us,
except that they lack the sort of thing
we think of as a voice.
We tell ourselves they are fainter
than we are, less defined,
that they are what we are defining,
that we are doing them a favour,
which makes us feel better.
They are less elegant about pain than we are.

But look, I said *us*. Though I may hate your guts
individually, and want never to see you,
though I prefer to spend my time
with dentists because I learn more,
I spoke of us as *we*, I gathered us
like the members of some doomed caravan

which is how I see us, travelling together,
the women veiled and singly, with that inturned
sight and the eyes averted,
the men in groups, with their moustaches
and passwords and bravado

in the place we're stuck in, the place we've chosen,
a pilgrimage that took a wrong turn
somewhere far back and ended
here, in the full glare
of the sun, and the hard red–black shadows
cast by each stone, each dead tree lurid
in its particulars, its doubled gravity, but floating
too in the aureole of *stone*, of *tree*,

and we're no more doomed really than anyone, as we go
together, through this moon terrain
where everything is dry and perishing and so
vivid, into the dunes, vanishing out of sight,
vanishing out of the sight of each other,
vanishing even out of our own sight,
looking for water.

A PAINTING OF ONE
LOCATION ON THE PLAIN

It's a journey with no end
in sight, and no end
after all; this place
is merely an oasis

which you can see as temporary,
a one-night stand,
if permanence makes you nervous.

There's nothing for us to do
apart from what is needed
to keep life going:
we eat, we drink, we lie down
together; if you want more,
there's the sky to admire,
serene as the evening before a war,
a few dry trees
and the vacancy of the desert

where tomorrow we will go out again
to make tracks in the sand
the wind will cover soon enough,
along with the bones
we also can't help making.

You could be sad because there isn't more,
or happy because there is
at least this much.

The sun's too hot, the water's poor,
the food is minimal, and I still
believe in free will:

the feast we're eating
in this golden haze of dust
is either a dead animal
or a blessing

or both. Take your life in your hands,
watch how it runs through your fingers
like sand or your own blood
into the ground you stand on

which is covered with stones and hard
and cloudy and endless as heaven.

Now you know where you are.

HEART TEST WITH AN ECHO CHAMBER

Wired up at the ankles and one wrist,
a wet probe rolling over my skin,
I see my heart on a screen
like a rubber bulb or a soft fig, but larger,

enclosing a tentative double flutter,
the rhythm of someone out of breath
but trying to speak anyway; two valves opening
and shutting like damp wings
unfurling from a grey pupa.

This is the heart as television,
a softcore addiction
of the afternoon. The heart
as entertainment, out of date
in black and white.
The technicians watch the screen,
looking for something: a block, a leak,
a melodrama, a future
sudden death, clenching
of this fist which goes on
shaking itself at fate.
They say: It may be genetic.

(There you have it, from science,
what God has been whispering all along
through stones, madmen and birds' entrails:
hardness of the heart can kill you.)

They change the picture:
now my heart is cross-sectioned
like a slice of textbook geology.
They freeze-frame it, take its measure.

A deep breath, they say.
The heart gasps and plods faster.
It enlarges, grows translucent,
a glowing stellar
cloud at the far end
of a starscope. A pear
made of smoke and about to rot.
For once the blood and muscle
heart and the heart of pure
light are beating in unison,
visibly.

Dressing, I am diaphanous,
a mist wrapping a flare.
I carry my precarious
heart, radiant and already
fading, out with me
along the tiled corridors
into the rest of the world,
which thinks it is opaque and hard.
I am being very careful.
O heart, now that I know your nature,
who can I tell?

THE SIDEWALK

We're hand in hand along
any old street, by the lake this time, and laughing
too at some joke we've
made and forgotten, and the sun
shines or it's raining, lunch after lunch, dinner
after dinner. You could see it
as one thing after another. Where
are we going? It looks like
nowhere; though we're going
where love goes finally, we're
going under. But not
yet, we're still
incarnate, though the trees break
into flame, blaze up, shed
in one gasp, turn to ash, each thing
burns over and over and we will
too, even the lake's
on fire now, it's evening and the sidewalk
fills with blue light, you can see down
through it, we walk on
water for a split
second before faith lapses and we let go
of each other also. Everything's
brighter just before, and it's
just before always.

THE WHITE CUP

What can I offer you, my hands held open,
empty except for my hands?

There is nothing to be afraid of,
you don't need my blessing.

As for the pigeons and the cedars
fading at dusk and emerging in early morning,
they can get along maybe
even better without me noticing them.

Coming back from a long illness
you can see how the white cup, the nasturtiums
on the porch, everything shines
not flagrantly as it did during your fever
but only the way it does.

This is the one thing I wanted to give you,
this quiet shining
which is a constant entering,
a going into

THE SKELETON,
NOT AS AN IMAGE OF DEATH

Your flesh moves under my fingers

and I remember *flesh* and *fingers*, as a child holding
the head of a flashlight cupped in my fist
in a dark room, seeing with such delight
the outlines of my own hand's
lucent skeleton, swathed in the red glow
of the blood clouded within

and this is how I hold
you: not as body,
as in planetary,
as in thing, bulk, object
but as a quickening,

a disturbance of the various
darknesses within my arms
like an eddy in the moonlit
lake where a fish moves unseen.

We rot inside, the doctor
said. To put a hand on another
is to touch death,
no doubt. Though there is also

this nebulous mist of interstellar
dust snagged by the gravity
of a few bones, mine,
but luminous:

even in the deep subarctic
of space beyond meaning, even among
the never alive, to approach
is to shine.

I hold you as I hold
water, swimming.

THE LIGHT

I see it as a light in a room.

The room itself is ordinary.
You can supply it yourself with whatever
you find most plausible: a table, a chair
with scuffed upholstery, a red vase, something
personal in bad taste. Wallpaper.
A plate, two plates
for dinner if you like. Maybe a bed.

The light is ordinary too,
it comes from a lamp, glass, with a fish on it
and a shade painted with rushes
and mallards, except that
there's no source for its energy.
No oil, no cord, no candle. It burns
on nothing. Or, it does not burn,
it shines merely. You open the door
and go into the room. The light is there.
You go out and close the door, you can no longer
see the light. Where does the light go when the door closes?
And you are outside and elsewhere.

Nowhere, you say. And that is
where the light shines endlessly, full
and inexhaustible. It shines nowhere.

THE BURNED HOUSE

At the centre of the burned space
in a clearing of red pines,
their weathered black scars towards me,
I lie in the full sun
on a wooden floor that will never be anything
but a floor.
Thirty years ago this was the room
I slept in, which is now wind.

I know the weather here
as the blind know colour. The sky
is motionless and heavy blue,
a cicada draws its little wiry
hot saw through the air.
In my fingers' ends there is thunder.

The grey floor glares,
the seared trees heat up,
the skin on my hands puckers.

They are not the same hands:
the former hands have grown out
layer by layer
and have been eaten away.

I stretch my new hands into the flames
which burned here and are still burning
slowly and unseen: that hesitation
which passes over the flesh
like breath riffling water,
that withering,
that shimmer

A STONE

On the wooden table the flame
of the lamp burns upwards without sound
and the smaller souls are called out by it
from where they have hidden during the day
in rotting stumps and the loosening
bark of trees
and hit softly against the window,
their feathery bellies licking the glass.

A loon ripples the night air
with its tune of clear silver,
the light blues,
and the one who has always been there
comes out of the shadows.

Have you had enough happiness? she says.
Have you seen
enough pain? Enough
cruelty? Have you had enough
of what there is? This
is as far as it goes.
Now are you ready for me?

Dark mother, whom I have carried with me
for years, a stone in my pocket,
I know the force of gravity
and that each thing is pulled downwards

against its will.
I will never deny you
or believe in you
only. Go back into your stone
for now. Wait for me.

SUMACS

*

At night I leave food out
on a white plate, milk in a white cup
and sit waiting, in the kitchen chair
beside last month's newspapers
and the worn coat leaning against
the wall: the shape of you left in the air,
time seeping out of it.
In the morning nothing has been touched
again. It is the wrong time of year.

*

Later, the days of the week unhook
from their names; the weeks unhook.
I do not lock the door
any more, but go outside and down
the bank, among the sumacs
with their tongues of dried blood
which have stopped speaking, to the pond
with its blackening water
and the one face wavering in it

wordless. Heaviness of the flesh infests me,
my skin that holds me in its nets;
I wish to change shape, as you have done
and be what you are

but that would be untrue also.

*

I lie on the damp yellowed grass bent
as if someone has been walking here
and press my head to the ground.

Come back, I tell you.
It becomes April.

If the daffodil would shed its paper
husk and fold back into its teardrop
and then down into the earth
into its cold onion
and into sleep. The one place I can still meet you.

*

Grief is to want more.
What use is moonlight?
I reach into it, fingers open,
and my hand is silvered
and blessed, and comes back to me holding nothing.

A BOAT

Evening comes on and the hills thicken;
red and yellow bleaching out of the leaves.
The chill pines grow their shadows.

Below them the water stills itself,
a sunset shivering in it.
One more going down to join the others.

Now the lake expands
and closes in, both.

The blackness that keeps itself
under the surface in daytime
emerges from it like mist
or as mist.

Distance vanishes, the absence
of distance pushes against the eyes.

There is no seeing the lake,
only the outlines of the hills
which are almost identical,

familiar to me as sleep,
shores unfolding upon shores
in their contours of slowed breathing.

It is touch I go by,
the boat like a hand feeling
through shoals and among
dead trees, over the boulders
lifting unseen, layer

on layer of drowned time falling away.

This is how I learned to steer
through darkness by no stars.

To be lost is only a failure of memory.

A BLAZED TRAIL

(i)

It was the pain of trees
that made this trail;
the fluid cut flesh of them only
partially hardened.
It is their scars that mark the way
we follow to the place where
the vista has closed over
and there is no more foresight.

(ii)

To blaze is also to burn.
All pathways through this burning
forest open in front of you and close
behind until you lose them.

This is the forest of lost things:
abandoned boulders. Burrows.
Roots twisted into the rock.
A toad in its cool
aura; an earthstar, splayed open
and leathery, releasing dust.
None of these things knows it is lost.

(iii)

We've come to a sunset, red and autumnal,
another burial. Although it is not
autumn, the wind has that chill.
Slight wind of a door closing.
The final slit of the old moon.

(iv)

I pick my way slowly
with you through the blazed forest,
scar by scar, back through
history, following the rule:

To recover what you have lost,
retrace your footsteps to the moment
at which you lost it. It will be there.

Here is the X in time.
When I am alone finally
my shadow and my own name
will come back to me.

(v)

I kneel and dig with my knifeblade
in the soil and find nothing.
I have forgotten what I hid here.

It must be the body of clear air
I left here carefully buried
and thought I could always
come back to and inhabit.

I thought I could be with myself only.
I thought I could float.
I thought I would always have a choice.
Now I am earthbound.
An incarnation.

(vi)

This is the last walk
I will take with you in your absence.

Your skin flares where I touch it,
then fades and the wood solidifies
around you. We are this momentary.

How much I love you.
I would like to be wise and calm.

I would make you eternal,
I would hold back your death if I could,
but where would you be without it?

We can live forever,
but only from time to time.

(vii)

Now we have reached the rocky point
and the shore, and the sky is deepening,
though the water still holds light
and gives it out, like fumes
or like fire. I wait, listening to that
place where a sound should be
and is not,
which is not my heart
or yours, which is darker
and more solitary,
which approaches. Which is the sound
the earth will make for itself
without us. A stone echoing a stone.
The pines rushing motionless.

INTERLUNAR

Darkness waits apart from any occasion for it;
like sorrow it is always available.
This is only one kind,

the kind in which there are stars
above the leaves, brilliant as steel nails
and countless and without regard.

We are walking together
on dead wet leaves in the intermoon
among the looming nocturnal rocks
which would be pinkish grey
in daylight, gnawed and softened
by moss and ferns, which would be green,
in the musty fresh yeast smell
of trees rotting, earth returning
itself to itself

and I take your hand, which is the shape a hand
would be if you existed truly.
I wish to show you the darkness
you are so afraid of.

Trust me. This darkness
is a place you can enter and be
as safe in as you are anywhere;
you can put one foot in front of the other
and believe the sides of your eyes.
Memorize it. You will know it
again in your own time.
When the appearances of things have left you,
you will still have this darkness.
Something of your own you can carry with you.

We have come to the edge:
the lake gives off its hush;
in the outer night there is a barred owl
calling, like a moth
against the ear, from the far shore
which is invisible.
The lake, vast and dimensionless,
doubles everything, the stars,
the boulders, itself, even the darkness
that you can walk so long in
it becomes light.